U.S. DEPARTMENT OF DEFENSE CONTRACT SPENDING AND THE INDUSTRIAL BASE, 2000–2013

Project Director
David J. Berteau

Lead Authors
Jesse Ellman
Gregory Sanders
Rhys McCormick

A Report of the CSIS National Security Program on Industry and Resources

October 2014

CSIS | CENTER FOR STRATEGIC & INTERNATIONAL STUDIES

ROWMAN & LITTLEFIELD
Lanham • Boulder • New York • Toronto • Plymouth, UK

About CSIS

For over 50 years, the Center for Strategic and International Studies (CSIS) has worked to develop solutions to the world's greatest policy challenges. Today, CSIS scholars are providing strategic insights and bipartisan policy solutions to help decisionmakers chart a course toward a better world.

CSIS is a nonprofit organization headquartered in Washington, D.C. The Center's 220 full-time staff and large network of affiliated scholars conduct research and analysis and develop policy initiatives that look into the future and anticipate change.

Founded at the height of the Cold War by David M. Abshire and Admiral Arleigh Burke, CSIS was dedicated to finding ways to sustain American prominence and prosperity as a force for good in the world. Since 1962, CSIS has become one of the world's preeminent international institutions focused on defense and security; regional stability; and transnational challenges ranging from energy and climate to global health and economic integration.

Former U.S. senator Sam Nunn has chaired the CSIS Board of Trustees since 1999. Former deputy secretary of defense John J. Hamre became the Center's president and chief executive officer in April 2000.

CSIS does not take specific policy positions; accordingly, all views expressed herein should be understood to be solely those of the author(s).

ISBN: 978-1-4422-4043-8 (pb); 978-1-4422-4044-5 (eBook)

Center for Strategic & International Studies
1616 Rhode Island Ave, NW
Washington, DC 20036
202-887-0200 | www.csis.org

Rowman & Littlefield
4501 Forbes Boulevard
Lanham, MD 20706
301-459-3366 | www.rowman.com

Contents

List of Figures and Table

Acknowledgments

This series of reports on the U.S. Department of Defense Contract Spending and the Industrial Base, now in its fourth edition, was initiated at CSIS in 2011. This report was only possible thanks to the dedicated efforts of numerous past staff members and interns as well as those that worked on this present report. Among our colleagues at the Center for Strategic and International Studies, we would like to thank Joshua Archer, Sam Brothers, Ryan Crotty, Madison Riley, Alex Stephenson, and Jing You for their considerable assistance in reviewing the text or preparing the data used in this series of studies. We would also like to thank our program coordinator Meredith Boyle and her predecessor Nicole Darden for keeping the program running during this effort. Additionally, we would like to thank Keith Tidman for his extremely detailed and through editing of this report, and Arwen McNierney for a focused look at the methodology section. Finally, we would like to thank all those in government that have answered our questions and whose hard work and dedication make this research possible.

Executive Summary: Defense Contracting Today

Over the past decade, the Center for Strategic and International Studies (CSIS) has issued a series of analytical reports on federal contract spending for national security and across the government. This latest report analyzes contracting for products, services, and research and development (R&D) by the U.S. Department of Defense (DoD) and its key components, using publicly available data from the Federal Procurement Data System (FPDS). It provides an in-depth look at current and recent trends in contract obligations as defense spending increased, then declined during the 2000–2013 study period. In particular, this report highlights the impact on DoD of sequestration in FY2013.

This edition expands on DoD reports from previous years through the use of supplemental material on the project website (http://www.csis.org/NSPIR/DoD). This material includes not only the graphs and tables contained in this report, along with the accompanying raw data, but also a variety of additional material arranged by defense component, product/service area, and contract/vendor characteristic. Throughout the year, the study team intends to publish and update the data as well as issue shorter publications on key issues relevant to the defense industrial base. This report, however, is independent of the website and stands on its own.

The first chapter of this report analyzes eight key facets of the defense contracting and the defense industrial base:

- By defense component
- By product/service area
- By level of competition
- By contract pricing mechanism
- By contract vehicle
- By size of contract
- By size of vendor
- Top 20 vendors

While most of these key facets are discussed in this executive summary, each is analyzed fully in subsequent chapters.

Methodology

Chapter 7 of the report describes the methodology used, including the study team's sources and methods, with a focus on changes in techniques and capabilities from prior reports. A few key methodological notes:

- All dollar figures are in constant 2013 billions.

- FPDS data are updated constantly as agencies revise submissions. Though the study team is confident in the overall trends, the precise dollar figures may continue to change (on the order of 1–2 percent) over the coming months.

- FPDS includes only prime contracts, and the separate subcontract database is not yet mature enough for CSIS analysis. Therefore, only prime contract data are included in this analysis.

- Reporting regulations only require that unclassified contracts be included in FPDS. The study team interprets this to mean that few, if any, classified contracts are reported into FPDS. For DoD, this omits a substantial amount of total contract spending, perhaps as much as 10 percent. Such omissions are probably most noticeable in R&D contracts.

- FPDS classifies contracts for research and development (R&D) in the broad category of services (as opposed to products). CSIS analyzes R&D separately from other services contracts, viewing DoD R&D spending as an investment more than as providing services.

Impact of Sequestration on Overall Defense Outlays and Funding Accounts

In Chapter 1, CSIS examines DoD contract obligations in the context of overall DoD outlays. Specifically, the study team looks at total defense gross outlays[1] and compares it with contract obligations. Under sequestration, gross defense outlays declined by -8 percent, falling from $702 billion in 2012 to $646 billion in 2013.

Contract obligations fell slightly faster than overall gross defense outlays

Between 2012 and 2013, with the impact of sequestration, defense-funded contract obligations[2] declined by 16 percent to $314 billion, a decline four times as steep as was seen during 2009–2012 budget drawdown. This cut was twice as steep as that of total defense gross outlays between 2012 and 2013. Meanwhile, noncontract gross outlays remained essentially flat from 2012 to 2013.

As a share of total gross defense outlays, defense funded contract obligations have declined from 53 percent to 49 percent in 2013, the lowest share since 2002.

---

[1] "Total defense gross outlays" includes outlays from three Office of Management and Budget (OMB) categories, excluding offsetting accounts: Department of Defense—Military Programs; Army Corps of Engineers—Civil Programs; and the portion of International Assistance Programs funded by DoD. See chapter 1 for more details.
[2] "Defense-funded contract obligations" includes all unclassified contracts in FPDS and funded by DoD, including those implemented by other agencies (such as by the Department of State). This total is, on average, approximately $10 billion higher than the total for contract obligations implemented by DoD. "Defense-implemented contract obligations" are used throughout this report except where specifically noted.

Overall, the Budget Control Act (BCA) of 2011 and the implementation of sequestration have had a significant effect on both overall defense outlays and defense contract obligations. The decline in total gross outlays under sequestration was comparable to that between 2011 and 2012, driven by the first tranche of defense spending reductions under the BCA. The decline in defense-funded contract obligations under sequestration, by contrast, was nearly three times as steep as between 2011 and 2012. The data show that a disproportionate share of the budget reductions under both the BCA and sequestration has been borne by contracts and by the defense industrial base.

This report for the first time tracks obligations by funding account

Chapter 1 of this report debuts analysis based on the availability of new government-wide data: the ability to track contract obligations by appropriations account for 2012 and 2013. These new data allow the study team to provide additional context on the effect of sequestration on defense contract obligations. Every appropriations account was hit hard by sequestration in 2013. Overall, defense-funded contract obligations declined more steeply than did the underlying appropriations accounts, including the procurement (-9 percent), research, development, test, and evaluation (RDT&E) (-10 percent),[3] and operations and maintenance (-15 percent) accounts. This was in part because the revolving and management funds account, primarily composed of various working capital funds (with DLA-associated working capital funds comprising around half), declined by 25 percent, though that figure is distorted by anomalies related to Defense Logistics Agency (DLA) fuels and the Defense Commissary Agency (DeCA) within that account. Defense-funded contract obligations within the foreign military sales (-20 percent), military construction (-37 percent), and military personnel and family housing (-19 percent) accounts all declined more rapidly than overall defense-funded contract obligations although the latter two represented a small drop in dollars.

Impact of Sequestration on Defense Contract Obligations in Specific Categories

Army and Navy back to prewar shares while DoD-wide agencies displace the Air Force

Overall defense contract obligations declined by 16 percent under sequestration, but that decline was not evenly distributed across the major DoD components, as discussed in greater detail in Chapter 2. Contract obligations for both the Army (-21 percent) and Air Force (-22 percent) declined more sharply under sequestration, driven by higher reductions in contract obligations for products and for R&D, whereas contract obligations for services were relatively preserved. This is the continuation of a trend for Army contract obligations, which declined by 15 percent in 2012. By contrast, Air Force contract obligations had increased by 4 percent in 2012, making the sharp decline under sequestration even more notable.

---

[3] Note that not all contract obligations classified by CSIS as R&D are appropriated out of the RDT&E account and that not all contract obligations in RDT&E are what CSIS classifies as R&D. See chapter 3 for more details.

Contract obligations by the Navy had declined by 12 percent in 2012 from 2011, twice that year's overall rate of decline in defense contract obligations. Under sequestration in 2013, however, Navy contract obligations declined by only 2 percent, one-eighth the rate of overall defense contract obligations. The category trends for Navy contract obligations are the opposite of those for the Army and Air Force: contract obligations for products increased, services declined disproportionately, and R&D declined the same as it had in previous years.

Cuts in DLA and "Other DoD" were largely driven by one-time factors

Contract obligations by the Defense Logistics Agency (DLA), which increased by 17 percent between 2011 and 2012, declined by 23 percent under sequestration. These figures are distorted by a large, anomalous one-year jump in DLA contract obligations for fuels in 2012, which subsequently returned to normal levels in 2013.

Contract obligations for the category of "Other DoD" (which includes all DoD contracting entities not covered by the other categories, including the Missile Defense Agency, TRICARE, and U.S. Transportation Command) declined by 18 percent under sequestration, after increasing by 2 percent in 2012. This trend is distorted by a reporting change: in 2013, the Defense Commissary Agency (DeCA) stopped reporting over $5 billion of annual nonappropriated funds contract obligations into FPDS.[4]

The impact of these two data anomalies for DLA and DECA is significant, as they account for more than two percentage points of the decline in defense contract obligations under sequestration.

Products declined faster than services, but this was largely due to a fuel spike in 2012

There is a similarly disparate trend when analyzing contract obligations broken down by what is being purchased. Contract obligations for products, which declined by only 3 percent in 2012 from 2011 (half the rate of overall defense contract obligations for that year), declined by 17 percent under sequestration. Excluding fuels, however, products declined by only 15 percent, and the DeCA reporting change accounted for as much as three percentage points of that decline.

Sequestration led to major declines in obligations for R&D contracts and for overall services contracts

R&D contract obligations, which declined by 11 percent in 2012, fell by an additional 21 percent under sequestration. This drop was 50 percent more than the average contract obligation decline in 2013. Contract obligations for services, which declined by 7 percent in 2012, declined another 14 percent in 2013, just slightly less than the decline in overall defense contract obligations under sequestration. Within the category of services, medical services (-4 percent), information and communications

---

[4] Though this is permitted under the Federal Acquisition Regulations (FAR), the study team nonetheless has concerns about the loss of visibility into over $5 billion of contract obligations. See chapter 2 for further discussion.

technology services (-8 percent), and professional, administrative, and management support services (-13 percent) fell less than the overall sequestration average, whereas equipment-related services (-19 percent) and facilities-related services and construction (-18 percent) saw larger-than-proportionate declines.

Defense R&D Contract Obligations under Sequestration

Though defense R&D contract obligations had been declining steadily for years (-8.1 percent compound annual growth rate (CAGR) between 2009 and 2012), the decline observed under sequestration was both quantitatively and qualitatively different (see Chapter 2 for more detail). Between 2012 and 2013, as overall DoD R&D contract obligations declined by 21 percent, Army (-35 percent) and Air Force (-27 percent) declined more steeply, whereas Navy (-10 percent) and "Other DoD" (-15 percent) fell less steeply during sequestration.

The decline in R&D contracts was not driven by the end of a handful of large contracts

Chapter 3 of this report examines possible root causes for these reductions in R&D contract obligations and concludes that, overall, the majority of the major declines in R&D contract obligations under sequestration were not tied to the cancellation or maturation into procurement account funding of major programs. Rather, cuts to more fundamental R&D, particularly in the missile and space realm, accounted for the largest share of R&D contract declines. Under sequestration, the major DoD components (particularly the Army and Air Force) were forced to make significant cuts in the stages of R&D that are critical to identifying and developing future critical technologies.

Middle-to-late stage R&D suffered the largest declines

Looking at R&D contract obligations by budget activity code under sequestration, basic research (-19 percent) and applied research (-18 percent) declined faster that overall DoD contract obligations, but less steeply than overall R&D. Mid-to-late-stage R&D (the 6.3 through 6.5 accounts) saw disproportionate declines: advanced technology development (-27 percent), advanced component development and prototypes (-24 percent), and systems development and demonstration (-21 percent) all declined more steeply than overall DoD. Meanwhile, contract obligations for operational systems development (-8 percent) declined less steeply under sequestration.

Final Thoughts, and the Difficulty of Prediction

This summary highlights some of the key impacts of sequestration on DoD contracting. The full report goes beyond sequester impacts and contains additional analysis of key contract and vendor characteristics. It also covers some of the key acquisition policy questions facing DoD, such as efforts to increase competition and to find the right balance in the promotion of fixed price contracting.

However, while the changes from 2012 to 2013 are significant in their own right, it is not yet clear whether they necessarily indicate continuing trends. As discussed throughout the paper, those categories hardest hit in 2013 were often those that took comparatively smaller cuts from 2011 to 2012. Volatility is to be expected until, one way or the other, planning and legal limits align. Nonetheless, the changes discussed in this report are meaningful in their own right, and in most cases these cuts are unlikely to be restored during a period of drawdown. What remains to be determined is whether DoD contract obligations will continue to decline faster than noncontract outlays and what the distribution of those declines will be across DoD components and contract categories.

1. Defense Budget under Sequestration

This chapter examines trends in defense contract obligations, both overall and broken down by major DoD component. To provide context for that analysis, the first section of this chapter places contract obligations in the context of overall defense outlays, to provide a basis for understanding the role of contracting in DoD.

Overall Defense Contract Obligations and Outlays

Figure 1-1: Overall Defense Contract Obligations and Outlays, 2000–2013

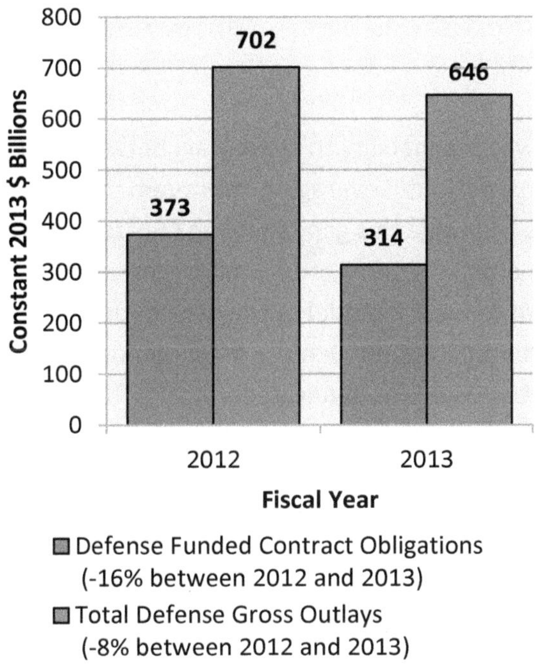

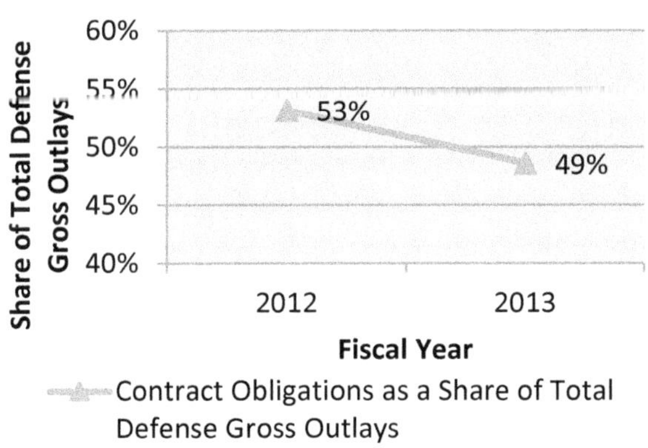

Source: FPDS; OMB historical data; CSIS analysis.

Under sequestration, contract obligations fell faster than overall outlays

Figure 1-1 shows defense-funded contract obligations (in blue) in the context of total defense gross outlays (in red). "**Defense-funded contract obligations**" includes all unclassified contracts *funded* by DoD, including those implemented by other agencies (such as the Department of State). This total is, on average, approximately $10 billion higher than the total for contract obligations *implemented* by DoD. Most of the tables and figures in this report represent contract obligations implemented by DoD, except where specifically noted. "**Total defense gross outlays**" includes outlays from three OMB categories, excluding offsetting accounts: Department of Defense–Military Programs; Army Corps of Engineers–Civil Programs; and the portion of International Assistance Programs funded by DoD. Both mandatory and discretionary outlays from these three agency categories are included, since some contracts (primarily for foreign military sales) are funded via mandatory spending accounts. Department of Defense–Civil Programs, which is primarily made up of offsetting accounts for retirement and healthcare, is excluded because it includes minimal contracting activity.

The green line, in the lower figure, shows defense-funded contract obligations as a share of total defense gross outlays. Note that the secondary axis starts at 40 percent.

Total defense gross outlays fell by 8 percent under sequestration. The main driver of this decline was the sharp decline (-16 percent) in defense-funded contract obligations under sequestration; excluding contracts, defense gross outlays increased by 1 percent between 2012 and 2013. During the 2009–2012 period, contract obligations had declined more modestly (-3.9 percent CAGR), and since their peak in 2009, contract obligations have declined by 25 percent.

As a share of total defense gross outlays, defense-funded contract obligations fell from 53 percent in 2012 to 49 percent in 2013. 2013 is the first year that contract obligations have accounted for less than half of total defense gross outlays since 2002.

How did sequestration affect defense contract obligations by appropriations account?

Prior to this report, a major limitation for the study team was the inability to cross-walk between FPDS contracting data and OMB/Treasury budget data and appropriations. Leveraging improvements in available data analysis tools, the study team has been able to bridge this gap, enabling CSIS to examine DoD contract obligations by Treasury account code, which enables CSIS to cross-walk between FPDS and OMB budget data. This, in turn, has allowed the study team to build a model of the major appropriations accounts, such as operations and maintenance (O&M) and procurement. See the Methodology section in Chapter 7 for a more detailed description of how this cross-walk was generated.

Going forward, CSIS plans to utilize this new capability in analysis of both defense contracting trends and broader defense budget trends. To preview this capability, the study team presents Figure 1-2, examining defense contract obligations by appropriations account.

Figure 1-2: Defense Contract Obligations by Appropriations Account, 2012 and 2013

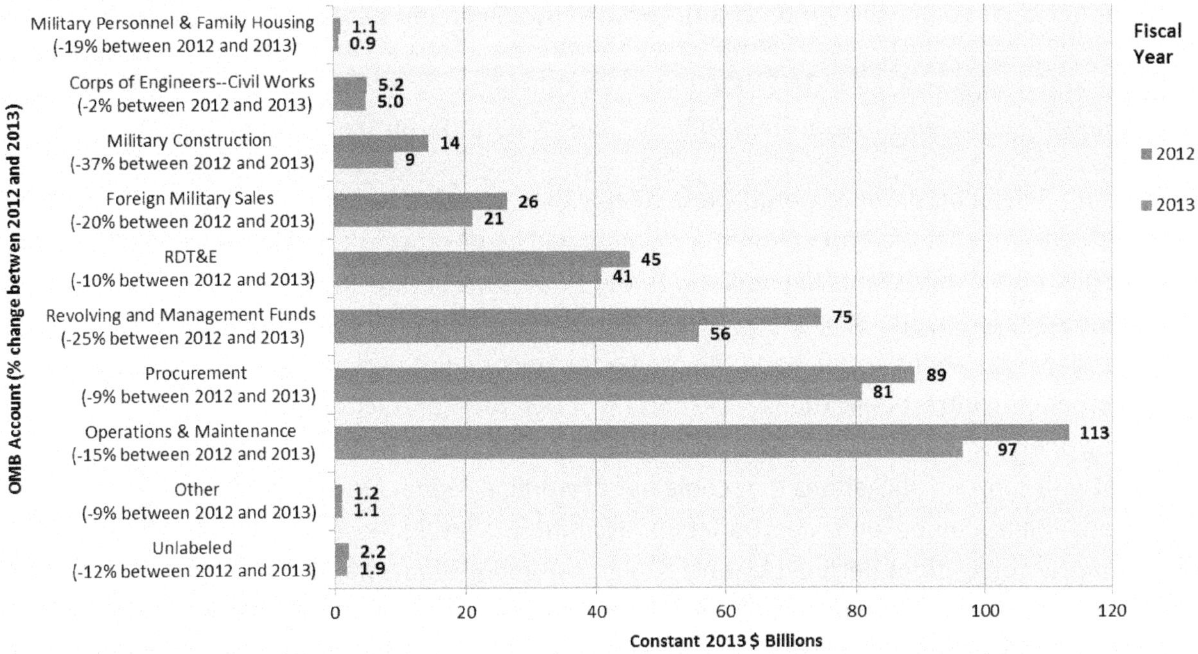

Source: FPDS; CSIS analysis.

This chart is limited to the past two years, as prior to 2012, the fields in the FPDS data necessary for the cross-walk (*Treasuryagencycode*, *mainaccountcode*, and *subaccountcode*) were largely left blank for DoD, precluding any meaningful trend analysis. Note also that these appropriations accounts do not line up with the study team's breakdowns of contract obligations by products/services/R&D, which depend on government Product or Service Codes (PSCs) for categorization of obligations. For example, only two-thirds to three-fourths of the contract obligations CSIS classify as R&D are in the research, development, test, and evaluation (RDT&E) account, and only around half of contract obligations under RDT&E are what CSIS classifies as R&D—the other major elements are products (likely early production funded out of RDT&E) and PAMS (likely R&D management support services that CSIS does not categorize as R&D).

RDT&E was hard hit, procurement was relatively protected, and against expectations O&M obligations were not disproportionately targeted

As overall defense-funded contract obligations[5] declined by 16 percent between 2012 and 2013, contract obligations within the O&M appropriations account declined slightly more slowly (-15 percent), which is surprising, given the conventional wisdom that O&M would be a major target for budget savings under sequestration. By contrast, the Revolving and Management Funds account, which, for

[5] As described with Figure 1-1, "defense-funded contract obligations" includes all contracts funded by DoD, including those implemented by other agencies (such as by the Department of State). This total is, on average, approximately $10 billion higher than the total for contract obligations implemented by DoD, which is the standard used for all other charts in this report.

contracting purposes, is primarily composed of various working capital funds (with DLA-associated working capital funds comprising around half), declined by 25 percent between 2012 and 2013.

The slower fall in procurement does not contradict the drop in product spending, because fuels and commissary purchases were most affected and are funded by other accounts

Procurement (-9 percent) declined notably less steeply than overall DoD contract obligations, defying widespread concerns that the Procurement account would be disproportionately targeted for budget savings. At first glance, this seems to conflict with the 17 percent decline in contract obligations for products described in Chapter 3, as there would seem to be significant overlap between the procurement appropriations account and the products category. But two of the major drivers of the decline in products contract obligations—the decline of DLA fuel contract obligations back to previous levels after a one-year jump and the removal of over $5 billion of DeCA contract obligations from FPDS—deal with contract obligations that come out of working capital funds, which fall under the Revolving and Management Funds appropriations account. Excepting those two sources of decline, products contract obligations declined by only 12 percent under sequestration, which is nearly in line with the level of decline seen in the Procurement appropriations account. Likewise, the DLA fuels decline and DeCA data removal account for $13 billion of the $19 billion decline in the Revolving and Management Funds account—putting aside those two sources of decline, the Revolving and Management Funds account fell by only 8 percent under sequestration.

The steepest declines in R&D spending take place outside of the RDT&E accounts

Contract obligations within the RDT&E account declined by 10 percent between 2012 and 2013, which is notably less steep than the decline in both overall defense and in R&D contract obligations as categorized by PSC (-21 percent). R&D contract obligations within the RDT&E account declined by 17 percent, whereas R&D contract obligations appropriated from non-RDT&E accounts declined by 29 percent. Professional, administrative, and management support (PAMS) contract obligations within RDT&E (largely for R&D management support services) declined by 22 percent, whereas products contract obligations under the RDT&E account increased by 7 percent, driven by large increases in contract obligations for fixed-wing aircraft and guided missiles.

Contract obligations under the Foreign Military Sales appropriations account (which is formally called the Military Sales Program) declined by 20 percent between 2012 and 2013, though this is likely more a factor of the cyclical nature of foreign military sales than the result of sequestration. Contract obligations under the Military Construction account (-37 percent) declined at over twice the rate of overall DoD, whereas Corps of Engineers–Civil Programs (-2 percent) were largely preserved, likely in part due to the impact of Hurricane Sandy. Contract obligations under the Military Personnel & Family Housing accounts declined at a rate (-19 percent) slightly faster than overall DoD under sequestration, but there was only around $1 billion in contract obligations under those accounts. Meanwhile, Other (-9 percent) and Unlabeled (-12 percent) declined more slowly than overall DoD.

Final Thoughts

This analysis represents a first look at the capabilities CSIS has developed to cross-walk between contract obligations, outlays, and appropriations. The study team will continue to explore new avenues of analysis made possible by these new capabilities, and would welcome feedback from our readers in regards to how these capabilities can be utilized to better inform the public debate on defense budget and contracting trends.

2. Defense Contract Obligations by Component

Figure 2-1: Overall Defense Contract Obligations by Component, 2000–2013

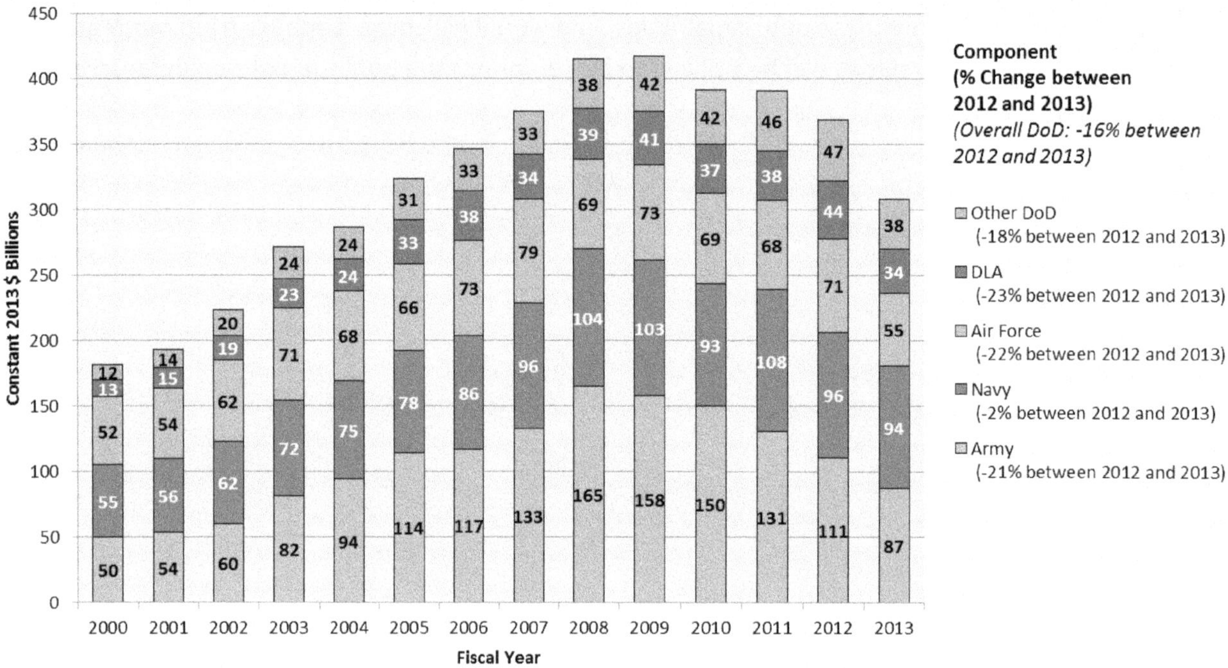

Source: FPDS; CSIS analysis.

Figure 2-1 looks at defense contract obligations by major DoD component, and then delves deeper to identify notable contracting trends within each component.

Army and Navy back to prewar shares while DoD-wide agencies displace the Air Force

Looking at the 2000–2013 period, the most significant trend was the growth in the share of DoD contract obligations awarded to the Army during the conflicts in Iraq and Afghanistan. The Army's share of contract obligations rose from 27 percent in 2002 to a high of 40 percent in 2008, but has fallen back to previous levels since, as U.S. ground forces have left Iraq and prepare to leave Afghanistan. Conversely, the share of DoD contract obligations awarded to the Navy declined steadily as Army rose, from 30 percent in 2000 to between 24 percent and 26 percent from 2004–2010, but has risen since back to previous levels. The Air Force, which accounted for 28 percent of DoD contract obligations from 2000–2002, has declined steadily since, and in 2008, the combined share of the Defense Logistics Agency (DLA) and "Other DoD" (comprising all contracting entities within DoD not covered by the other four components, such as the Missile Defense Agency (MDA), TRICARE, and U.S. Transportation Command) exceeded the share for the Air Force for the first time. This trend has continued in every year since.

Many of the decisions on how to comply with the sequestration budget limits were made at the component level within DoD, so it is important to examine and compare how the major DoD components fared. Contract obligations by the Army, which declined by -15 percent between 2011 and 2012 and at an -11.2 percent CAGR between 2009 and 2012, declined even more dramatically in 2013 (-21 percent). Air Force contract obligations, which were mostly steady (-0.9 percent CAGR between 2009 and 2012) during the drawdown, and which increased by 5 percent between 2011 and 2012, declined even faster than for the Army (-22 percent). Contract obligations by DLA declined at the highest rate (-23 percent), but this comes after a large one-year surge in contract obligations for fuels in 2012. A large portion of that decline from 2012 is simply a return to prior levels of obligations for fuels.

Interestingly, Navy contract obligations, which declined at over twice the rate of overall DoD between 2011 and 2012 (-11 percent), and declined at a -2.4 percent CAGR between 2009 and 2012, declined at one-eighth the rate of overall DoD between 2012 and 2013 (-2 percent). The Navy saw significant declines in contract obligations for services and R&D, but those were mostly balanced by increases in contract obligations for products.

"Other DoD" largely avoids declines, with the exception of a commissary accounting change

The category of "Other DoD" declined only slightly faster than overall defense contract obligations between 2012 and 2013 (-18 percent), after showing mild growth (3 percent) between 2011 and 2012 and a similar rate of growth (3.5 percent CAGR) between 2009 and 2012. The main driver of this decline is the disappearance of most contract obligations by the Defense Commissary Agency (DeCA) from FPDS between 2012 and 2013, with reported contract obligations falling from $6.7 billion in 2012 to $1.5 billion in 2013. Under Federal Acquisition Regulation (FAR) 4.606(b)(4), resale activity (such as for commissaries) is not required to be reported into FPDS; evidently, DeCA has decided to stop reporting those contract obligations that they are not required to report into FPDS. In conversations with DoD policymakers, they were aware of this decision and did not oppose it, since it concerned nonappropriated funds. Since this appears to be a permanent issue going forward, CSIS will explore removing DeCA contract obligations from the back year datasets in future editions of the report, to allow for more meaningful comparisons.

To better understand the contracting environment within DoD under sequestration, the following subsections analyze notable contracting trends within the major DoD components.

Army

Notable cuts in weapon systems, R&D, medical services

As overall Army contract obligations declined by 21 percent between 2012 and 2013, Army contract obligations for products declined by 27 percent, double the rate of decline from 2009–2012 (-13.4 percent CAGR). The most notable declines were a nearly $1 billion decline in contract obligations for drones, a nearly $900 million decline in contract obligations for combat assault and tactical vehicles, and an over $500 million decline in contract obligations for land mines.

Army R&D contract obligations declined steeply (-35 percent), at nearly triple the rate of decline from 2009–2012 (-13.5 percent CAGR). See Chapter 3 for more information on the drivers of this decline in R&D contract obligations.

Army services contract obligations declined more slowly (-15 percent) than overall Army. Within services, information and communications technology (ICT) services (-4 percent) declined at less than a fourth the rate of overall Army services, whereas medical (MED) services (-27 percent) declined at nearly twice the rate of overall Army services.

A precipitous decline in time and materials contracts

Contract obligations under time and materials contract types, which had been declining rapidly from 2009–2012 (-21.8 percent CAGR), fell at more than double that rate (-46 percent) between 2012 and 2013. Contract obligations under cost reimbursement contract types, which had held steady during the drawdown (-0.2 percent CAGR) even as overall Army dropped sharply, declined by 9 percent under sequestration, still less than half the rate of overall Army.

Navy

In 2013, Navy shifted toward buying more products, principally the Joint Strike Fighter

As overall Navy contract obligations declined by only 2 percent between 2012 and 2013, Navy products contract obligations increased by 8 percent, several times the rate of growth from 2009–2012 (1.3 percent CAGR). Notable sources of growth included the Joint Strike Fighter program ($7.4 billion), nuclear reactors ($1.1 billion), the H-1 Upgrade program ($800 million), CVN-68 ($800 million), DDG-51 ($750 million),[6] and the E-2C Advanced Hawkeye ($500 million). Several programs saw significant reductions, including the LHA(R) (-$2.1 billion), the P-8 Poseidon[7] (-$1 billion), and the V-22[8] (-$900 million). Navy R&D contract obligations, which had been declining at a -10.2 percent CAGR from 2009–2012, fell at virtually the same rate under sequestration.

Navy services contract obligations, which declined moderately over the 2009–2012 drawdown period (-4.7 percent CAGR), declined at almost four times that rate (-16 percent) under sequestration. Within services, professional, administrative, and management support (PAMS) services declined by 23 percent,

---

[6] For CVN-68, DDG-51, and other surface ship programs, the Navy seems to be consolidating type-specific PSC codes (such as "Aircraft Carriers" and "Destroyers") to a more general, catch-all PSC code ("Combat Ships and Landing Vessels"). The study team is concerned about this reduction in data granularity and will monitor the issue going forward.

[7] Prior to 2013, most of the contract obligations for the P-8 Poseidon were classified by system equipment code under the old program title (Multimission Maritime Aircraft, MMA).

[8] Oddly, the Navy seems to have reclassified the V-22 as a fixed-wing aircraft, shifting most of the V-22 contract obligations from the "Aircraft, Rotary Wing" PSC code to the "Aircraft, Fixed Wing" PSC code between 2012 and 2013.

driven by a $1.2 billion decline for "Special Studies/Analysis–Technology." Facilities-related services and construction (FRS&C) declined by 21 percent, whereas equipment-related services (ERS) declined by only 2 percent.

An increasing share for weapon systems also leads to less competition, more reliance on the Big 6 defense vendors, and more definitive contracts. The share of Navy contract obligations awarded without competition continued its upward trend since 2010, from 54 percent in 2012 to 59 percent in 2013, tied with the Air Force for the highest of any of the major components. Competition with 5+ offers declined at 11 times the rate (-22 percent) of overall Navy contract obligations under sequestration, and over three times the rate during the 2009–2012 drawdown (-7.3 percent CAGR).

There was a significant shift in the Navy's use of contract pricing mechanisms, as the share of contract obligations awarded under fixed price contract types rose from 58 percent in 2012 to 63 percent in 2013, whereas cost reimbursement fell from 37 percent to 32 percent. A similar trend was observed in the Navy's use of contract vehicles: the share contract obligations awarded under definitive contracts rose from 62 percent to 67 percent, whereas single award indefinite delivery contracts (IDCs) fell from 20 percent to 16 percent. Both changes continued trends observed since 2010.

Navy contract obligations awarded to the Big 6 defense vendors rose by 14 percent under sequestration, increasing their share of overall Navy contract obligations from 39 percent to 45 percent. Large vendors, which had been growing at a 4.5 percent CAGR from 2009–2012, declined by 11 percent between 2012 and 2013. Small vendors, which had been declining at a -1.9 percent CAGR from 2009–2012, fell by 13 percent between 2012 and 2013. These shifts reflect the Navy's apparent focus on preserving contract obligations for its largest programs under sequestration, most of which are carried out by vendors in the Big 6.

Air Force

Air Force trends invert that of Navy due to cutbacks in the C-17, the EELV, and fixed-wing aircraft

As overall Air Force contract obligations declined by 22 percent between 2012 and 2013, Air Force products contract obligations, which had been increasing at a 3.0 percent CAGR from 2009–2012, fell by 28 percent. The main drivers of the decline are cuts related to the C-17A (-$3.5 billion), the Evolved Expendable Launch Vehicle (EELV) (-$2.3 billion), and a $3 billion decline for uncategorized fixed-wing aircraft that the study team believes to be related to the F-35 Joint Strike Fighter.[9] Air Force services

[9] By system equipment code, the Air Force has less than $1 million in contract obligations categorized under the F-35 in 2012 and 2013 combined, an obvious data error resulting from the system equipment code field being left blank. Oddly, the data also show a $900 million increase in contract obligations for fixed-wing aircraft categorized under the system equipment code for the Shillelagh missile, an Army antitank missile program from the 1960s and

contract obligations, which had been declining slightly faster than overall Air Force from 2009–2012, declined less steeply than overall Air Force in 2013 (-15 percent). Within services, ICT showed slight growth (1 percent) after declining sharply during the 2009–2012 budget drawdown (-14.0 percent CAGR). Air Force R&D contract obligations, which had declined moderately (-4.3 percent CAGR) between 2009 and 2012, fell by 27 percent under sequestration. See Chapter 3 for more information on the drivers of this decline in R&D contract obligations.

Drop in products leads to a higher percent competed and a decline in Big 6 market share

The share of Air Force contract obligations awarded without competition, which had risen steadily since 2005, declined from 63 percent in 2012 to 59 percent in 2013. Contract obligations awarded under fixed price contract types, which had been increasing at an 8.3 percent CAGR during the 2009–2012 budget drawdown, fell by 25 percent under sequestration. Similarly to the Army, Air Force contract obligations under time and materials contract types fell dramatically (-65 percent).

Contract obligations awarded to the Big 6 defense vendors, which had been growing slightly (1.2 percent CAGR) during the 2009–2012 budget drawdown, fell by 31 percent under sequestration, likely due to the previously mentioned cuts in contract obligations for large aircraft procurement programs. Contract obligations awarded to medium contractors, who had seen moderate growth (6.3 percent CAGR) during the 2009–2012 period, declined by 13 percent under sequestration, which was still notably less steep than the decline for overall Air Force.

Defense Logistics Agency

DLA returns to its mean, after a notable spike in fuel purchases in 2012

As overall DLA contract obligations declined by 23 percent, nonfuels products, which had been increasing modestly (2.4 percent CAGR) between 2009 and 2012, fell by 13 percent. Contract obligations awarded after competition with 5+ offers, which spiked between 2011 and 2012, declined by 35 percent, returning to previous levels. As a share of overall DLA contract obligations, competition with 5+ offers fell from 60 percent in 2012 to 50 percent in 2013. The share of contract obligations awarded after competition with two offers (7 percent in 2012, 12 percent in 2013) and three to four offers (10 percent in 2012, 13 percent in 2013) rose in parallel. In 2013, 99.7 percent of DLA contract obligations were awarded under fixed price contract types.

Contract obligations awarded to large vendors declined by 35 percent under sequestration, falling as a share of overall DLA contract obligations from 49 percent in 2012 to 41 percent in 2013. Medium vendors, which had seen moderate growth (6.3 percent CAGR) during the 2009–2012 budget

1970s. CSIS urges Air Force policymakers to emphasize greater attention to proper FPDS data entry for the system equipment code data field, reinforced by the publishing of a system equipment code book that avoids code reuse.

drawdown, declined by 10 percent under sequestration, but still grew as a share of overall DLA contract obligations, from 28 percent in 2012 to 33 percent in 2013.

Other DoD

Commissary purchases are delisted and air and sea freight contracting drops notably

As overall "Other DoD" contract obligations declined by 18 percent between 2012 and 2013, largely driven by the disappearance of $5 billion in contract obligations by DeCA, services declined notably less steeply (-12 percent). Within services, ERS, which had grown at a 31.5 percent CAGR between 2009 and 2012, declined by over half (-51 percent) under sequestration, driven by a $1.6 billion drop in contract obligations for "Air Charter of Things," and a $900 million decline in contract obligations for vessel freight. Contract obligations for MED were nearly steady (-1 percent), and PAMS increased by 8 percent, over double the rate of growth from 2009–2012 (3.9 percent CAGR). The main driver of the increase in PAMS contract obligations was a $600 million increase in contract obligations for "Passenger Air Charter Service."

"Other DoD" contract obligations awarded without competition, which had been growing slightly (2.3 percent CAGR) from 2009–2012, declined by 27 percent under sequestration, falling as a share of overall "Other DoD" contract obligations from 28 percent in 2012 to 24 percent in 2013. Contract obligations awarded under fixed price contract types, which had been growing strongly (8.7 percent CAGR) during the 2009–2012 budget drawdown, fell by 32 percent, declining as a share from 55 percent to 45 percent. Contract obligations under cost reimbursement contract types held nearly steady between 2012 and 2013 (1 percent), rising as a share from 43 percent to 54 percent.

Contract obligations awarded under definitive contract types increased by 5 percent under sequestration, rising as a share from 41 percent to 52 percent, while all categories of IDCs fell by about a third. Contract obligations awarded to medium vendors, which had been growing moderately (6.1 percent CAGR) from 2009–2012, declined by 34 percent under sequestration, falling as a share from 29 percent in 2012 to 23 percent in 2013. Meanwhile, contract obligations awarded to the Big 6 defense vendors increased by 5 percent, rising as a share from 19 percent to 24 percent.

Final Thoughts on Components

The CSIS analysis in this chapter reflects selective samples of data and assessments. Additional data, including breakdowns of contract and vendor trends by major DoD component, are available at http://www.csis.org/NSPIR/DoD. CSIS will update both the data and the analysis in response to new policy developments, or as additional data analysis capabilities become available.

3. Products, Services, and R&D

Defense Contract Obligations by Area

Figure 3-1: Overall Defense Contract Obligations by Area, 2000–2013

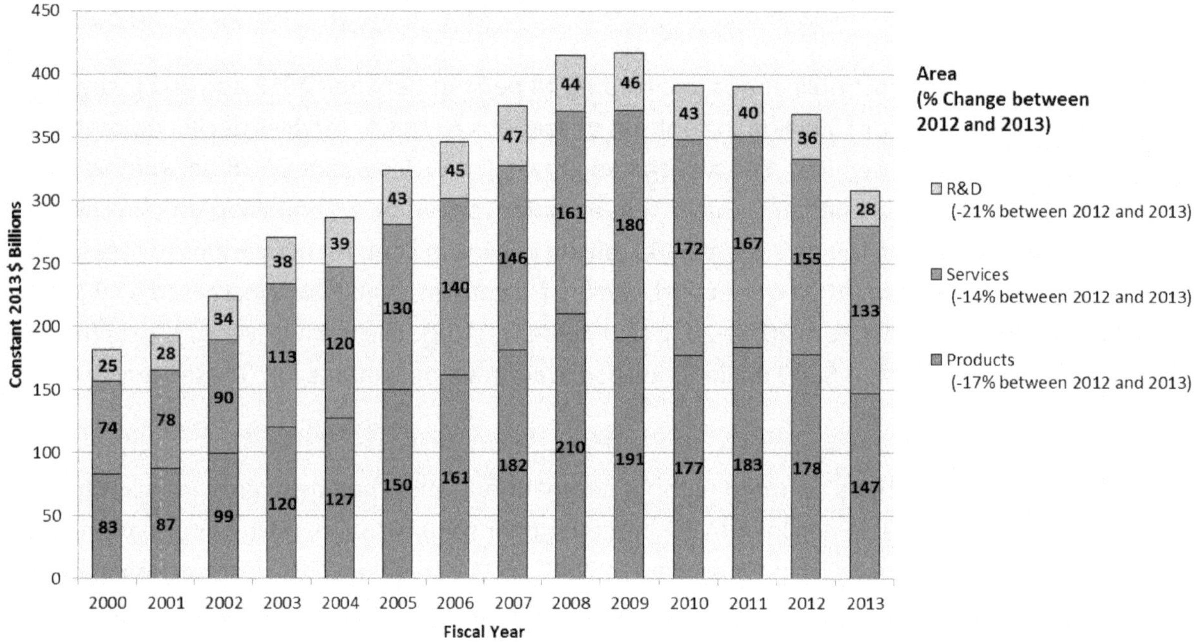

Source: FPDS; CSIS analysis.

Figure 3-1 looks at defense contract obligations by what is being purchased, and then delves deeper into notable contracting trends within each area.

From 2000–2013, the shares of DoD contract obligations awarded for products and services have been remarkably consistent. Products accounted for between 44 percent and 48 percent of contract obligations in all but two years between 2000 and 2013 (49 percent in 2007, 51 percent in 2008), and services have accounted for between 39 percent and 43 percent of contract obligations in all but one year (44 percent in 2010) during the period. By contrast, the share of defense contract obligations going to R&D has declined steadily, from a high of 15 percent in 2002 to a low of 9 percent in 2013. In absolute terms, DoD in 2013 obligated less for R&D contracts than in any year since 2001.

Obligations for services have declined at the same rate of overall spending, even when targeted

There have been numerous anecdotal reports, both in the media and in conversations that the study team has had with industry, about how contract obligations for services were being specifically targeted for cuts under sequestration. Indeed, the rate of decline under sequestration is twice the rate of decline for services between 2011 and 2012 (-7 percent) and almost three times the rate of decline for the entire 2009–2012 drawdown period (-4.9 percent CAGR). However, as seen in Figure 3-1, however, contract obligations for services have actually declined more slowly (-14 percent) than overall DoD between 2012 and 2013. Within services, contract obligations for ICT (-8 percent) and MED (-4 percent) both declined more slowly than overall services contract obligations; both had declined more slowly than overall services during the 2009–2012 period. PAMS, which have been targeted for cuts in recent years (in particular through OMB guidance to reduce contract obligations for management support services), but had declined at a rate comparable to overall DoD contract obligations and identical to overall DoD services from 2009–2012 (-4.9 percent CAGR), declined similarly to overall services under sequestration (-13 percent). FRS&C had been declining sharply between 2009 and 2012 (-11.7 percent CAGR), and that trend continued in 2013 (-18 percent). And ERS, which was the only service area to see increases between 2009 and 2012 (5.5 percent CAGR), declined the most of any service area in 2013 (-19 percent).

Products declined faster than services, but this was largely due to a fuel spike in 2012

Due to a decline in contract obligations for fuels, products declined at a rate slightly faster than for overall DoD (-17 percent) under sequestration, after declining more slowly than overall DoD between 2009 and 2012 (-2.4 percent CAGR). This was driven by a sharp decline in contract obligations for fuels by DLA after an anomalous one-year increase, as discussed in Chapter 2. The decline in contract obligations for products was also inflated by the removal of $5 billion in products contract obligations under DeCA from FPDS, also discussed in Chapter 2. Accounting for the fuels obligations anomaly and the DeCA data issue, contract obligations for products declined (between -12 and -13 percent) significantly less steeply than did overall defense.

Sequestration led to major declines in obligations for R&D that cannot be traced to changes in a small number of large major defense acquisition programs

Contract obligations for R&D declined notably faster than overall DoD contract obligations (-21 percent), after declining at twice the rate of overall DoD contract obligations between 2011 and 2012 (-10 percent), and twice the rate of overall DoD between 2009 and 2012 (-8.1 percent CAGR). This conforms to the predictions of Secretary Frank Kendall in a September 2013 speech: "If we're talking about an overall budget cut of 10 percent across the department, you can about double that for the R&D and procurement accounts."[10] Unlike the 2009–2012 period, where the declines in R&D were driven by major defense acquisition programs (MDAPs) either being canceled or maturing out of R&D,[11] the decline between 2012 and 2013 appears to be broad-based and not primarily tied to a few large development programs. These data seem to indicate that DoD has exhausted its ability to preserve overall R&D spending by targeting large, underperforming programs or letting programs mature from

---

[10] Jared Serbu, "DoD research, acquisition face double whammy from sequestration," Federal News Radio, September 5, 2013, http://www.federalnewsradio.com/394/3442329/DoD-research-acquisition-face-double-whammy-from-sequestration.

[11] See the previous version of this report, Gregory Sanders et al., *U.S. Department of Defense Contract Spending and the Supporting Industrial Base, 2000–2012* (Washington, DC: CSIS, December 2013), 23–24, http://csis.org/files/publication/131207_Sanders_DODContractSpending_Web.pdf.

R&D to procurement, and that sequestration has led to real, significant cuts in DoD investment in research and development.

To better understand the contracting environment within DoD under sequestration, the following subsections analyze notable contracting trends for products, services, and R&D. (Trends for products/services/R&D within the major DoD components are discussed in the previous section.)

Products

The end of a spike in DLA fuel contracts leads to a shift away from competition with 5+ offers

As overall DoD products contract obligations declined by 17 percent under sequestration, products contract obligations awarded after competition with 5+ offers declined at over twice that rate (-37 percent), after having declined only marginally (-2.3 percent CAGR) from 2009–2012. Competition with 5+ offers fell as a share of overall DoD products from 19 percent to 15 percent. Contract obligations awarded without competition declined at a rate (-15 percent) comparable to that of overall DoD products, and increased as a share from 59 percent in 2012 to 61 percent in 2013.

Contract characteristics were shaped by continuing major weapon system contracts and decline in other types of products. Products contract obligations awarded under definitive contract types declined by only 5 percent between 2012 and 2013, whereas single-award IDCs declined by 33 percent. As a share, definitive contracts increased from 47 percent to 54 percent, whereas single-award IDCs declined from 39 percent to 32 percent (down from 46–47 percent between 2005 and 2010). Products contract obligations awarded to the Big 6 defense vendors declined by only 1 percent, growing as a share of DoD products contract obligations from 36 percent in 2012 to 43 percent in 2013. Large vendors saw a 25 percent decline, with their share falling from 32 percent to 29 percent. And medium vendors, which had held nearly steady (0.5 percent CAGR) during the 2009–2012 budget drawdown, declined by 31 percent under sequestration, falling as a share from 19 percent to 16 percent.

Services

Services contracts were not disproportionately cut, except for time and materials contracts

As overall DoD services contract obligations declined by 14 percent under sequestration, services contract obligations awarded under time and materials contract types, which had been declining sharply (-24.7 percent CAGR) during the 2009–2012 budget drawdown, fell by nearly half (-47 percent). The Big 6 defense vendors, which had largely maintained their services contract obligations during the 2009–2012 period (-0.9 percent CAGR), saw a 19 percent decline in services contract obligations under sequestration.

R&D

A decline in R&D of 21 percent coincides with greater shares for competition and cost-plus contracts

As overall R&D contract obligations declined by 21 percent, R&D contract obligations awarded without competition declined by 32 percent, falling as a share from 41 percent in 2012 to 36 percent in 2013. R&D contract obligations awarded after competition with two offers grew in parallel, rising as a share from 22 percent to 27 percent. The share of R&D contract obligations awarded under cost reimbursement contract types rose from 76 percent in 2012 to 84 percent in 2013.

Sequestration reversed the trend toward more fixed-price contracting for R&D. In the FY2012 edition of this report, the study team identified a notable rise in the use of fixed price contract types for R&D between 2009 and 2012 (19.9 percent CAGR). This trend reversed itself under sequestration, as fixed price declined by 44 percent, falling as a share of overall R&D contract obligations from 21 percent to 15 percent. For context, the share of R&D contract obligations awarded under fixed price contract types hovered between 7 percent and 10 percent in all but one year between 2003 and 2009, but the 2013 share is in line with rates seen between 2000 and 2002. Given the concerns expressed by both industry and DoD policy leaders about the possible overuse of fixed price contract types for programs where it was not appropriate, the study team will continue to monitor this issue in future editions of this report.

What are the specific sources of decline in DoD R&D contract obligations?

Unlike FY2012, this decline was not driven by the end of a handful of large contracts

Between 2012 and 2013, as overall DoD R&D contract obligations declined by 21 percent, Army (-35 percent) and Air Force (-27 percent) declined more steeply than overall DoD, whereas "Other DoD" (-15 percent) and Navy (-10 percent) managed to relatively preserve their R&D contract obligations during sequestration. In the previous edition of this report, the study team explored the sources of decline in DoD R&D contract obligations between 2009 and 2012. The main conclusion was that the primary driver of the decline was the cancellation or maturation from R&D into procurement account funding of specific large major defense acquisition programs (MDAPs), such as the Army's Future Combat Systems (FCS), rather than portfolio-wide reductions. For this report, the study team investigated to determine whether that root cause was still driving reductions in R&D contract obligations.

Army and "Other DoD" R&D cuts were driven reductions in applied/exploratory R&D accounts. and Missile Defense Agency support

For the Army, where nearly the entire decline in R&D contract obligations between 2009 and 2012 was due to the cancellation and gradual winding-down of FCS, the decline between 2012 and 2013 does not seem to be tied to any specific MDAP. The two main drivers of the decline in Army R&D contract obligations are an approximately $500 million decline in Missile Defense Agency support for advanced development of missile/space systems and an approximately $550 million decline in applied/exploratory

R&D not classified under any major acquisition program in FPDS. Together, these two cuts account for nearly half of the decline in overall Army R&D contract obligations.

Similarly, for "Other DoD," the main source of decline for R&D contract obligations under sequestration was an over $500 million reduction in contract obligations for Missile Defense Agency support (specifically, for basic research, applied research/exploratory development, and engineering development).

Air Force and Navy R&D cuts came from a range of sources and were partially offset by increases in the Advanced Extremely High Frequency satellite program and the F-35 respectively

For the Air Force, the picture is more mixed. From 2009–2012, Air Force R&D contract obligations declined at half the rate of overall DoD R&D contract obligations, a sharp contrast to the steep decline under sequestration. Between 2012 and 2013, there was an approximately $550 million increase in applied research/exploratory development tied to the Advanced Extremely High Frequency (AEHF) satellite program, but overall contract obligations declined by over $3 billion. The main drivers of the decline between 2012 and 2013 are more varied than with the Army:

- The disappearance of R&D contract obligations for the Wideband Gapfiller Satellite program (from $1.2 billion in 2012 to -$2 million (a de-obligation) in 2013)
- A $200 million decline in commercialized aircraft research relating to the F-22
- A nearly $500 million decline in basic research and engineering development related to uncategorized electronics/communications equipment
- A nearly $600 million decline in advanced development and applied research/exploratory development related to uncategorized missile and space systems
- An over $300 million decline in uncategorized applied research/exploratory development

For the Navy, moderate increases in R&D contract obligations related to the F-35 Joint Strike Fighter (JSF) and the Mobile User Objective System (MUOS) satellite program occurred, along with a $500 million reduction in R&D contract obligations for the E-2C Advanced Hawkeye.

Middle-to-late-stage R&D suffered the largest declines

Looking at R&D contract obligations by budget activity code under sequestration, basic research (-19 percent) and applied research (-18 percent) declined faster than overall DoD contract obligations, but less steeply than overall R&D. Mid-to-late-stage R&D saw disproportionate declines: advanced technology development (-27 percent), advanced component development and prototypes (-24 percent), and systems development and demonstration (-21 percent) all declined more steeply than overall DoD. Meanwhile, contract obligations for operational systems development (-8 percent) were relatively preserved under sequestration.

Overall, it appears the majority of the major declines in R&D contract obligations under sequestration were not tied to the cancellation or maturation into procurement account funding of MDAPs, but rather cuts to more fundamental R&D, particularly in the missile and space realm. DoD was able to largely protect R&D contract obligations during the 2009–2012 budget drawdown by targeting underperforming programs and shifting maturing programs into procurement. These data indicate that, under sequestration, the major DoD components (particularly the Army and Air Force) were unable to continue that trend, and were forced to make significant cuts in the stages of R&D that are critical to identifying and developing future critical technologies.

Final Thoughts on Products, Services, and R&D

The CSIS analysis in this chapter reflects selective samples of data and assessments. Additional data, including breakdowns of contract and vendor trends by products, services, and R&D, are available at http://www.csis.org/NSPIR/DoD. CSIS will update both the data and the analysis in response to new policy developments, or as additional data analysis capabilities become available.

4. Competition

Defense Contract Obligations by Competition

Figure 4-1: Overall Defense Contract Obligations by Level of Competition, 2000–2013

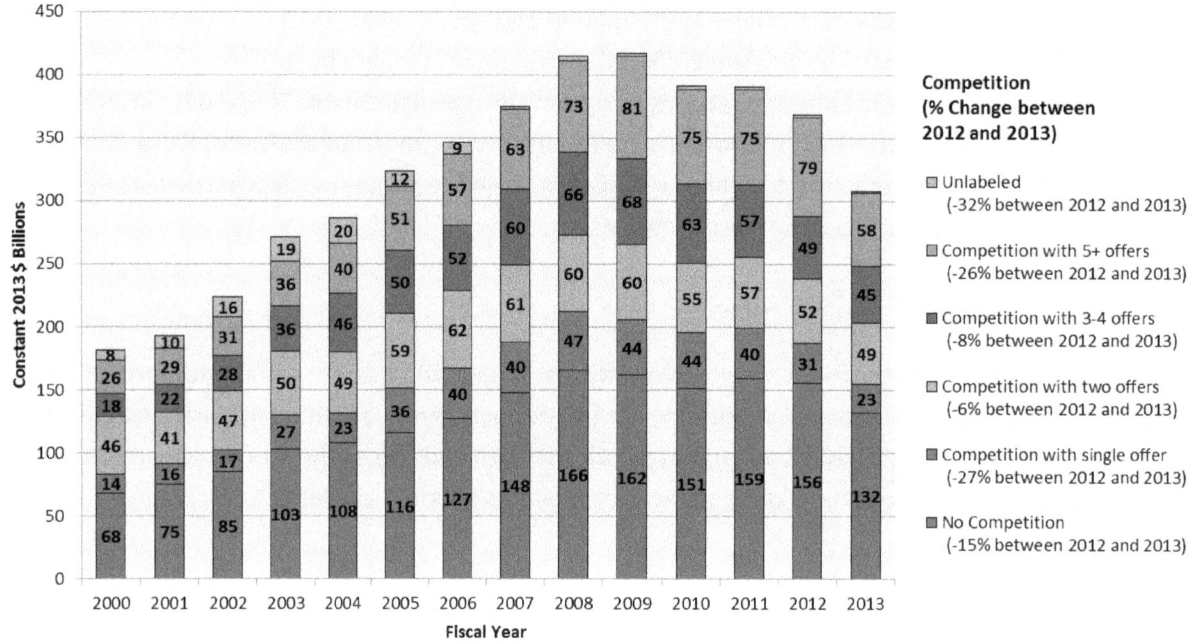

Source: FPDS; CSIS analysis.

Figure 4-1 examines the level of competition in defense contract obligations from 2000–2013, using number of offers as the primary determinant of competitiveness. The study team defines effective competition as a competition that receives two or more offers.

Effective competition accounts for about half of obligations, rising slightly as a share in 2013, whereas single-offer competition lost ground in recent years to no competition

The share of defense contract obligations awarded without competition has remained relatively steady throughout the 2000s, with between 37 percent and 40 percent of contract obligations awarded without competition in all but one year (36 percent in 2005), though that rate has risen slowly since 2010. Competed contracts that receive only one offer, which the study team has long cited as a prime target for improvements in competitive sourcing, grew from 8 percent of contract obligations in 2000 to between 10 percent and 11 percent from 2005–2011, but have declined since. Competition with two offers has declined steadily throughout the 2000–2013 period, from 25 percent in 2000 to between 14 percent and 16 percent from 2008–2013. The share of contract obligations awarded after competition with three to four offers grew from 10 percent in 2000 to between 15 percent and 16 percent in all but one year from 2004–2013 (13 percent in 2012). The share of contract obligations awarded after competition with five offers has grown similarly, from 15 percent in 2000 to a high of 21 percent in 2012. Overall, the rate of effective competition in DoD contracting has remained near 50 percent in most years. Unlabeled, which accounted for as much as 7 percent of overall defense contract obligations in the early 2000s, is no longer a significant issue.

Levels of competition in DoD contracting did not change significantly under sequestration. The share of contract obligations awarded without competition increased from 42 percent in 2012 to 43 percent in 2013, continuing a gradually increasing trend since 2010. The rate of effective competition rose from 48 percent in 2012 to 50 percent in 2013: competition with two offers and with three to four offers increased, while competition with five or more offers declined. Contract obligations awarded after competition with a single offer declined from 8 percent in 2012 to 7 percent in 2013, the lowest share in the 2000–2013 period.

How successfully have the different DoD components implemented guidance to increase competition in contracting?

Figure 4-2: Contract Obligations Awarded after Effective Competition, by Component, 2000–2013

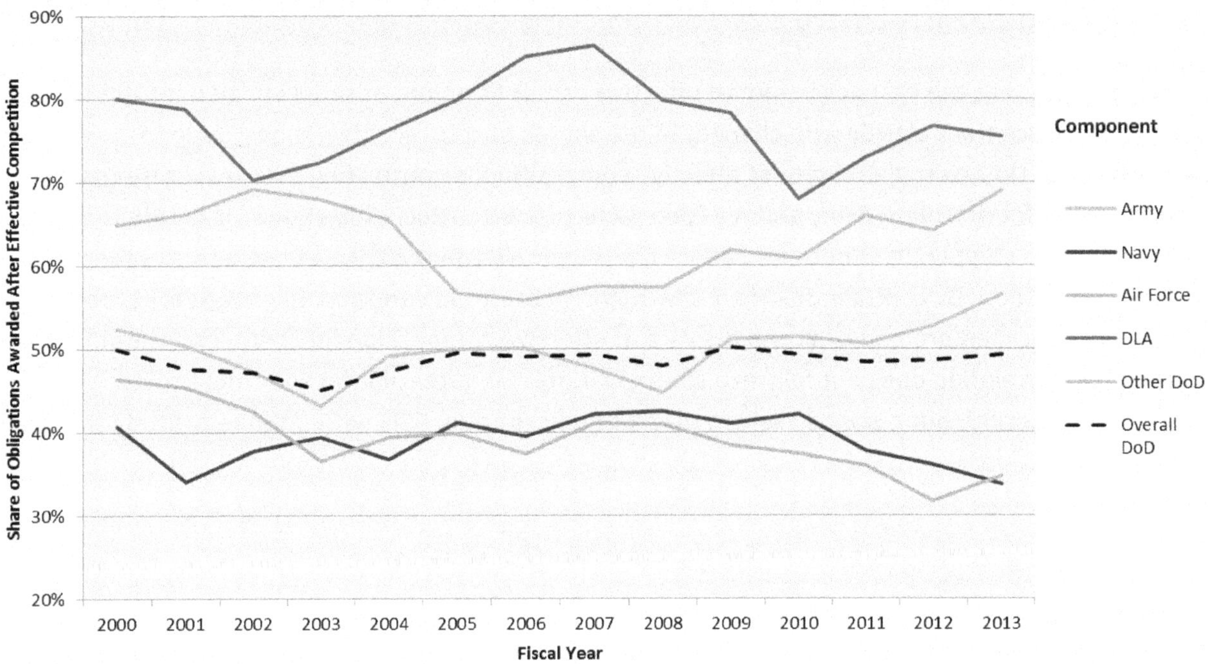

Source: FPDS; CSIS analysis.

Though levels of competition for overall DoD contract obligations have been relatively consistent, there are significant differences in how well major DoD components have implemented guidance to increase competition in contracting. Figure 4-2 breaks down the competition data further, examining the shares of contract obligations awarded after effective competition (that is, competition with two or more offers). The dotted black line shows the rate of effective competition for overall DoD between 2000 and 2013. The rate of effective competition for overall DoD was remarkably consistent, fluctuating between 47 percent and 50 percent in all but one year during the period (45 percent in 2003).

Navy and Air Force both have below average effective competition, whereas Army is average

Both the Navy and Air Force have had effective competition rates that have been significantly lower than overall DoD throughout the 2000–2013 period. The share of Navy contract obligations awarded after effective competition remained close to 40 percent through most of the 2000s, but has declined steadily since, from 42 percent in 2010 to 36 percent in 2012 to 34 percent in 2013. The share of Air Force contract obligations awarded after effective competition was 46 percent in 2000, but quickly declined, remaining around 40 percent from 2002–2008. Effective competition has declined steadily within Air Force contracting since then, from 41 percent in 2008 to 32 percent in 2012, before rising to 35 percent in 2013.The rate of effective competition for the Army has tracked closely to that of overall DoD, not differing by more than three percentage points until the last two years. The share of Army contract obligations awarded after effective competition remained near 50 percent in all but two years between 2000 and 2011 (43 percent in 2003, 45 percent in 2008), but has increased since, from 51 percent in 2011 to 53 percent in 2012 to 57 percent in 2013.

Army, Navy, and Air Force tend to have less competition when they buy more products

It is worth noting that the changes in rate of effective competition under sequestration for the Army, Navy, and Air Force track closely with changes in the shares of contract obligations going to products and services. For DoD overall, the rate of effective competition in contracting for services is notably higher than that for products. Army and Air Force, which saw contract obligations for products decline significantly more steeply than those for services between 2012 and 2013, saw increases in their rate of effective competition. By contrast, the Navy saw slight increases in contract obligations for products while services declined dramatically, and the rate of effective competition for Navy contract obligations declined. This correlation does not fully explain the variation in rates of effective competition throughout the period, but it seems likely to play a significant role. CSIS will continue to study this issue in the future.

DoD-wide agencies have high rates of competition, in part because they buy fuels, as well as transportation and medical services

Both DLA and "Other DoD" had rates of effective competition notably higher than those of overall DoD. This trend is likely a factor of what DLA and "Other DoD" primarily contract for: things like fuels, transportation services, and medical services make up a large share of contracting done by the two components. There is a broader market from which to draw competitors for those sorts of products and services than there is for some of the larger, more complex programs run by the Army, Navy, and Air Force. The share of "Other DoD" contract obligations awarded after effective competition declined from a peak of 69 percent in 2002 to 56 percent in 2006, rose back to 66 percent in 2011, declined to 64 percent in 2012, and increased to 69 percent in 2013. For DLA, the share of contract obligations awarded after effective competition has fluctuated significantly: from 80 percent in 2000 to 70 percent in 2002, up to 86 percent by 2007, back down to 68 percent by 2010, up to 77 percent in 2012, and then falling back to 75 percent in 2013.

Final Thoughts on Competition

The CSIS analysis in this chapter reflects selective samples of data and assessments. Additional data, including breakdowns of competition trends by major DoD component and area, are available at http://www.csis.org/NSPIR/DoD. CSIS will update both the data and the analysis in response to new policy developments, or as additional data analysis capabilities become available.

5. Defense-Industrial Vendor Base

Defense Contract Obligations by Vendor Size, 2000–2013

Figure 5-1: Overall Defense Contract Obligations by Vendor Size, 2000–2013

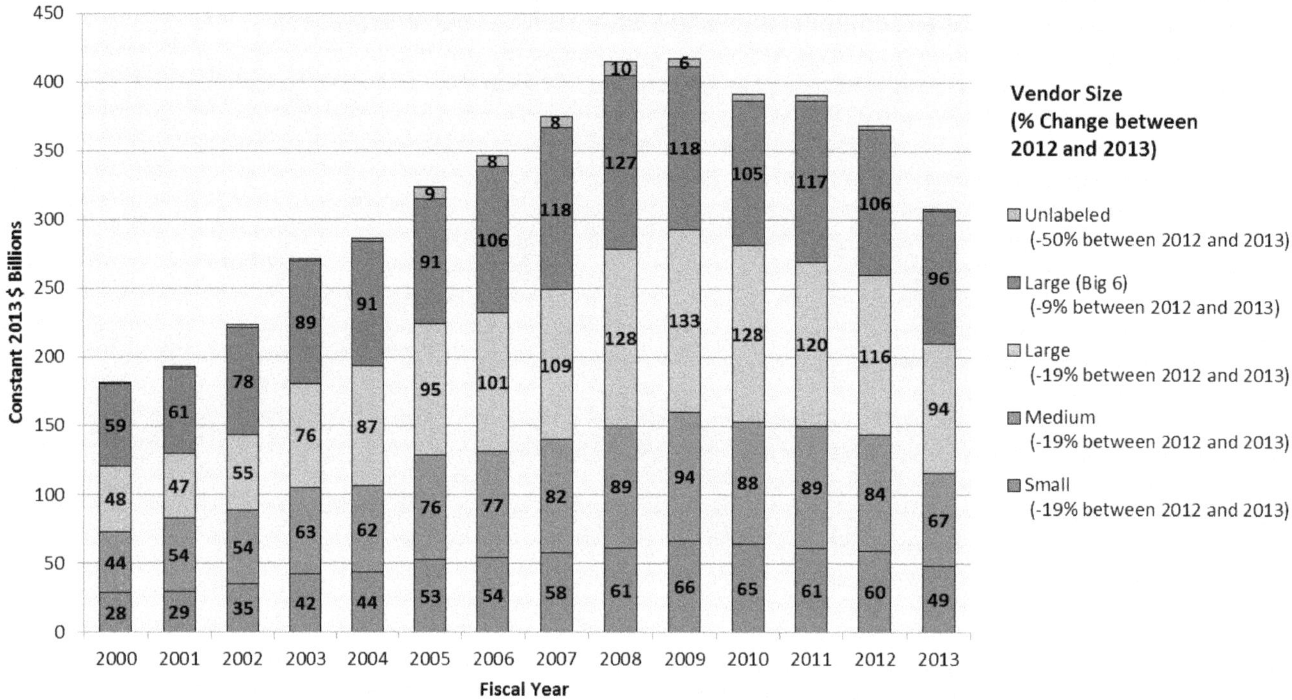

Source: FPDS; CSIS analysis.

Figure 5-1 examines defense contract obligations by size of vendor. To analyze the breakdown of competitors in the market into small, medium, and large vendors, the CSIS team assigned each vendor in the database to one of four size categories. Any organization designated as small by the FPDS database—according to the criteria established by the federal government—was categorized as such unless the vendor was a known subsidiary of a larger entity. Vendors with annual revenue of more than $3 billion, including from nonfederal sources, are classified as large. The Big 6 is a subset of "large," separating out the six largest defense firms (Boeing, Lockheed Martin, Northrop Grumman, General Dynamics, Raytheon, and BAE). And any contractor that qualifies as neither small nor large is classified as "medium."

Vendor shares have been consistent, although the Big 6 lost share after 2003 and have begun to regain it in recent years as $500 million-plus contracts faced less pressure from sequestration

Since the early 2000s, the shares of contract obligations awarded to the four size categories of vendors have been remarkably consistent. Small contractors have received between 15 percent and 16 percent of contract obligations in every year from 2000–2013. Medium contractors have received between 21 percent and 23 percent of contract obligations in every year from 2003–2013. Large contractors have received between 28 percent and 32 percent of contract obligations in every year since 2003. Only the Big 6 defense firms show a notable trend over the period: after accounting for 32 percent to 35 percent of contract obligations from 2000–2004, the Big 6 have accounted for only between 27 percent and 31 percent in the years since. Unlabeled, which grew to account for as much as 3 percent of overall DoD contract obligations in 2005, has declined steadily since, and is no longer a significant factor.

From 2009–2012, contract obligations in all four size categories declined at rates comparable to that of overall DoD contract obligations. Under sequestration, contract obligations awarded to small, medium, and large vendors all declined more steeply (-19 percent for each) than overall DoD contract obligations. Meanwhile, contract obligations for the Big 6 defense vendors declined at just over half the rate (-9 percent) of overall DoD. Given the context of the subsequent section on contract size, where the data showed that the largest contracts ($500 million or more in annual obligations) were relatively protected under sequestration, it makes sense that the Big 6 defense firms would have fared better under sequestration, as they account for a disproportionate share of the largest DoD contracts.

Top 20 Defense Vendors, 2003 and 2013

Table 5-1: Top 20 Defense Vendors, 2003 and 2013

Rank	Top 20 Contractors in 2003	Obligations in 2013 Millions	2002 Rank	Top 20 Contractors in 2013	Obligations in 2013 Millions	2012 Rank
1	Lockheed Martin	28,202	1	Lockheed Martin	36,972	1
2	Boeing	22,462	2	Boeing	19,898	2
3	Northrop Grumman	14,241	3	Raytheon	12,770	3
4	General Dynamics	10,556	5	General Dynamics	11,967	4
5	Raytheon	10,167	4	Northrop Grumman	9,859	5
Subtotal for Top 5		**85,628**			**91,465**	
6	United Technologies	5,663	6	L3 Communications	5,492	7
7	Halliburton	5,434	38	Huntington Ingalls	5,481	9
8	Stewart & Stevenson	5,180	39	United Technologies	5,339	6
9	L3 Communications	3,790	23	BAE Systems	4,961	8
10	SAIC	3,590	7	SAIC	4,331	10
11	General Electric	3,578	12	Humana	3,460	11
12	BAE Systems	3,520	11	Dyncorp International	3,006	17
13	Humana	3,017	13	Health Net	2,908	13
14	Health Net	2,236	9	ITT	2,871	27
15	Computer Sciences Corp.	1,947	17	Bechtel	2,758	20
16	ITT	1,842	18	Textron	2,656	25
17	URS	1,691	21	General Electric	2,231	22
18	Dyncorp	1,567	15	General Atomics	2,125	23
19	Honeywell	1,548	14	Bell-Boeing Joint Project Office*	2,096	14
20	TriWest Healthcare	1,528	22	Fluor	2,094	30
Total for Top 20		**131,760**			**143,272**	
Total for all industry		**270,957**			**307,974**	

Source: FPDS; CSIS analysis. *Note: May not sum due to rounding.*

* Joint Venture.

Table 5-1 shows the top 20 defense vendors in both 2003 and 2013, measured by prime contract obligations. The columns to the right of the dollar totals show the rank of that vendor in the previous year.

The market has been stable, although Northrup Grumman's divestiture of Huntington Ingalls Industries has led to changed position and reinforced a slight decentralization

Unsurprisingly, the composition of the top 5 defense vendors has not changed dramatically between 2003 and 2013. The only major shift is the decline by Northrop Grumman, which is mostly due to the divestiture of its shipbuilding business into Huntington Ingalls Industries, which ranked 7th in 2013. The remainder of the top 20 was also relatively stable, with 9 of the remaining 15 companies in the top 20 in 2003 still in the top 20 in 2013. Similarly, of the vendors in the top 20 in 2013, only five were outside of the top 20 in 2012.

There has been a small but noticeable de-concentration of the defense industrial base since 2003. As a share of the top 20, the top 5 vendors declined from 65 percent in 2003 to 64 percent in 2013. As a share of total defense contract obligations, the top 5 have declined from 32 percent in 2003 to 30 percent in 2013. Similarly, as a share of total defense contract obligations, the top 20 has declined from 49 percent in 2003 to 47 percent in 2013. In all three cases, those shares have increased by between two and four percentage points since 2012. This is consistent with the relatively small declines in contract obligations to the Big 6 defense vendors and contract obligations for contracts of $500 million or more noted previously.

Final Thoughts on the Defense-Industrial Vendor Base

The CSIS analysis in this chapter reflects selective samples of data and assessments. Additional data, including breakdowns of vendor size and top 10 vendor lists by major DoD component and area, are available at http://www.csis.org/NSPIR/DoD. CSIS will update both the data and the analysis in response to new policy developments, or as additional data analysis capabilities become available.

6. Contract Pricing, Vehicle, and Size

Defense Contract Obligations by Contract Pricing Mechanism

Figure 6-1: Overall Defense Contract Obligations by Contract Pricing Mechanism, 2000–2013

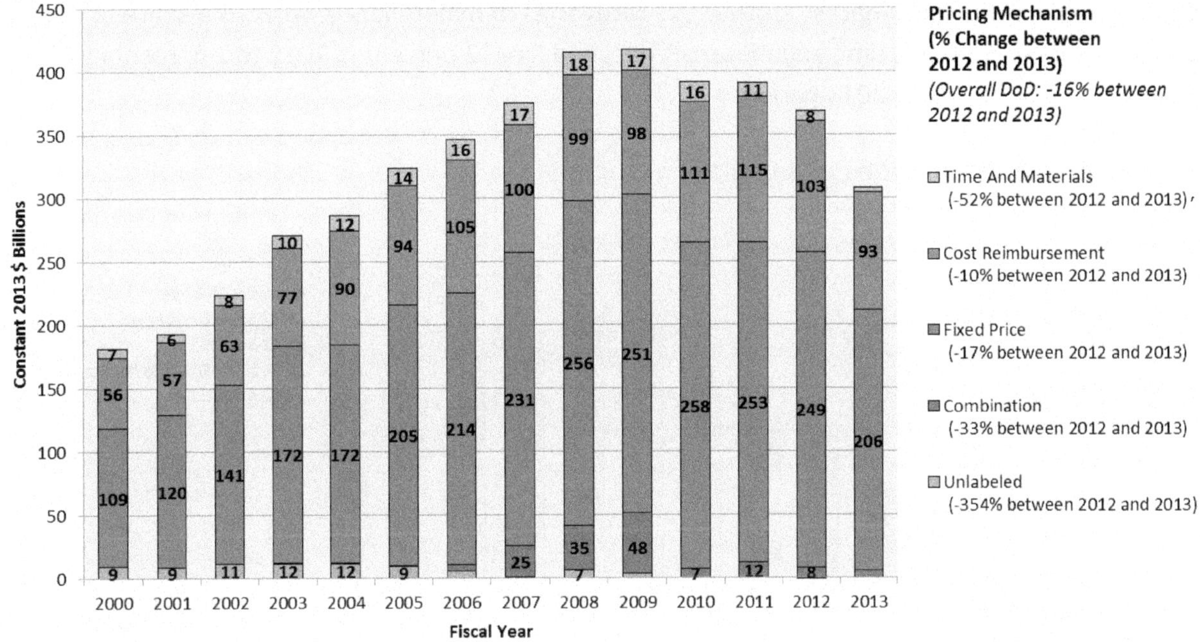

Source: FPDS; CSIS analysis.

Figure 6-1 looks at defense contract obligations by contract pricing mechanism.

The balance between cost reimbursement and fixed price contracts has been steady, although a slight recent trend toward fixed price contracts was reversed under sequester

DoD's use of contract pricing mechanism was remarkably consistent for most of the 2000s. Between 2000 and 2009, between 60 percent and 63 percent of DoD contract obligations were awarded under fixed price contract types. Cost reimbursement contract types accounted for between 28 percent and 31 percent of defense contract obligations from 2000–2006, with that share in the late 2000s dropping only because true levels were obscured by the significant rise of contract obligations categorized as "combination" from 2007–2009. Time and materials contract types accounted for between 3 percent and 5 percent from 2000–2011. And unlabeled has not accounted for even 1 percent of defense contract obligations since 2009. Contract obligations under fixed price contract types, which had nearly held steady (-0.3 percent CAGR) during the 2009–2012 budget drawdown, declined at a rate (-17 percent) comparable to that of overall DoD under sequestration. As a share of overall defense contract obligations, fixed price declined from 68 percent in 2012 to 67 percent in 2013. Cost reimbursement, which had been growing slightly from 2009–2012 (1.8 percent CAGR), declined at a rate (-10 percent) slower than overall DoD under sequestration, growing as a share of overall DOD from 28 percent to 30 percent.

Use of time and materials contracts has been dramatically reduced in recent years

Contract obligations under time and materials contract types, which had been declining sharply (-23.4 percent CAGR) during the 2009–2012 budget drawdown, fell by over half (-52 percent) under sequestration. The main sources of this decline were within the Army and Air Force, primarily for "Engineering and Technical Services" and various categories of maintenance work. As a share of overall defense contract obligations, time and materials declined from 2 percent in 2012 to 1 percent in 2013, the lowest share in the 2000–2013 period.

Defense Contract Obligations by Contract Vehicle

Figure 6-2: Overall Defense Contract Obligations by Contract Vehicle, 2000–2013

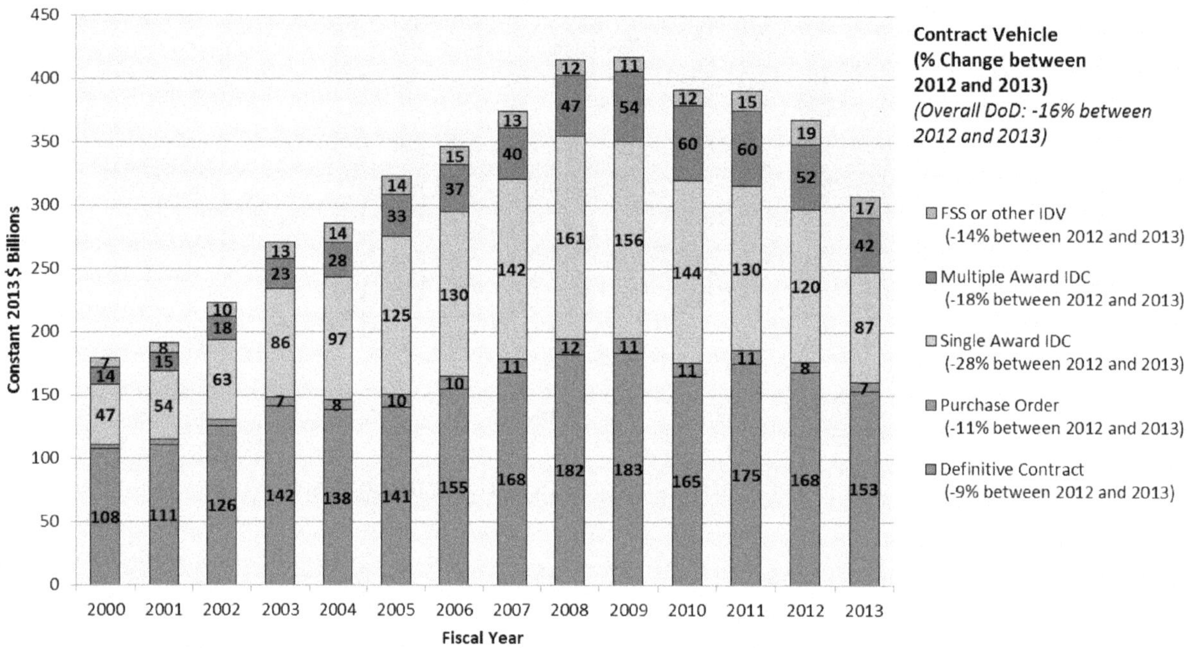

Source: FPDS; CSIS analysis.

Figure 6-2 looks at defense contract obligations by contract vehicle.

Single-award and multiple-award IDCs grew dramatically during the 2000s, but both have lost ground in recent years, particularly the former

The use of contract vehicles in DoD contracting has changed significantly in the 2000–2013 period. Definitive contracts, which accounted for 59 percent of DoD contract obligations in 2000, declined steadily throughout the 2000s, to a low of 42 percent in 2010, before rebounding in recent years. By contrast, the share of contract obligations awarded under single-award IDCs grew steadily, from 26 percent in 2000 to a high of 39 percent in 2008, before declining in recent years. The use of multiple-award IDCs had nearly doubled, from 8 percent in 2000 to 15 percent in 2010 and 2011, in line with FAR guidance to increase the use of multiple-award vehicles. Purchase orders (between 2 percent and 3 percent) and FSS or other IDVs (between 3 percent and 5 percent) have remained steady through most of the 2000–2013 period.

Definitive contracts regained market share as major weapon systems disproportionately avoided sequestration cuts

Definitive contracts, which had declined (-2.8 percent CAGR) more slowly than overall DoD during the 2009–2012 budget drawdown, fell by 9 percent under sequestration, just over half the rate of overall DoD. As a share of overall DoD contract obligations, definitive contracts increased from 46 percent in 2012 to 50 percent in 2013. Single-award IDCs, which had been declining significantly (-8.3 percent CAGR) between 2009 and 2012, declined at nearly three and a half times that rate (-28 percent) between 2012 and 2013, falling as a share from 33 percent to 28 percent.

Multiple-award IDCs, which had declined only slightly during the 2009–2012 budget drawdown (-1.8 percent CAGR), fell slightly faster (-18 percent) than overall DoD under sequestration, holding steady with a 14 percent share. FSS or other IDVs, which had been growing strongly (19.3 percent CAGR) from 2009–2012, declined sharply (-14 percent) between 2012 and 2013, albeit at a rate slightly slower than overall DoD. As a share, FSS and other IDVs held steady at 5 percent between 2012 and 2013. Purchase orders declined under sequestration (-11 percent) at nearly the same rate they had been during the 2009–2012 budget drawdown (-10.1 percent CAGR), holding steady with a 2 percent share in 2012 and 2013.

Defense Contract Obligations by Contract Size, 2000–2013

Figure 6-3: Overall Defense Contract Obligations by Contract Size, 2000–2013

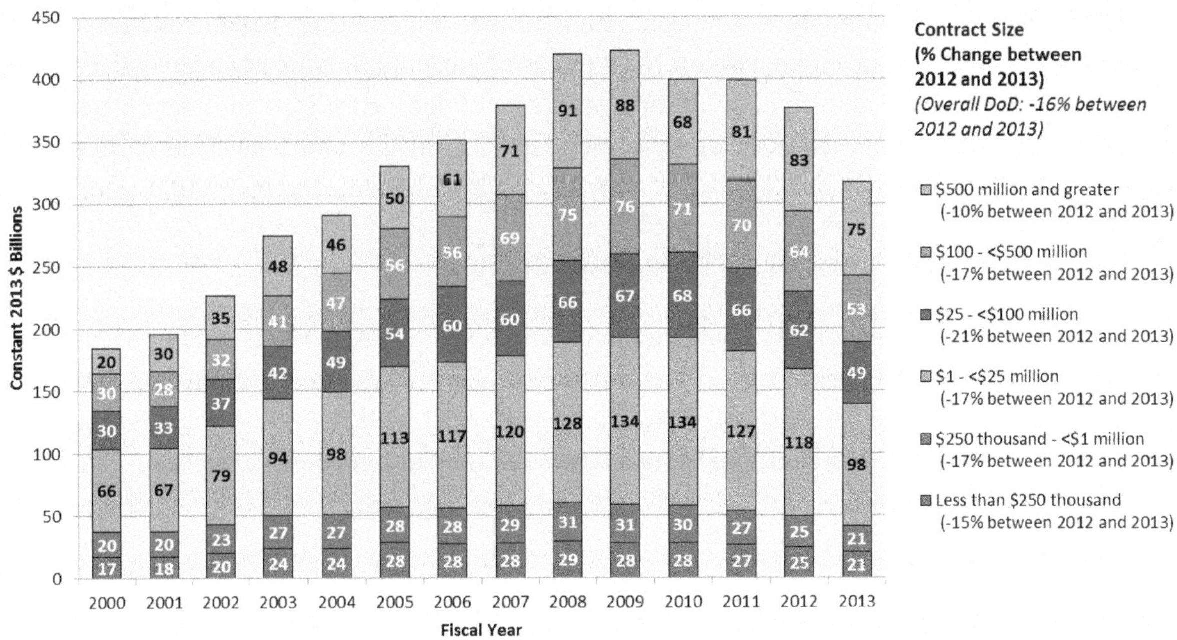

Source: FPDS; CSIS analysis.

Figure 6-3 looks at defense contract obligations by size of contract. For this analysis, contract size is defined by total annual appropriations under a contract in a given year. This chart also excludes de-obligations, so the totals differ from the other charts in this section.

The smallest contracts have been losing share whereas the largest have been gaining

The relative distribution of contract size for DoD contract obligations has changed over time, with share redistributed from the smallest contracts to the largest, whereas the shares awarded under medium-sized to large contracts have been remarkably consistent. The share of contract obligations for contracts under $250 thousand has declined steadily from 2000–2013, from 9 percent in the early 2000s to 7 percent from 2008–2013. Similarly, the share awarded under contracts obligating between $250 thousand and $1 million has declined from 11 percent in 2000 to 7 percent from 2011–2013. Meanwhile, the share awarded under contracts obligating $500 million or more has increased fairly steadily, from 11 percent in 2000 to 24 percent in 2013.

Contract obligations for contracts between $1 million and $25 million have fluctuated between 32 percent and 36 percent in all but one year from 2001–2013 (31 percent in 2008). Contracts obligating between $25 million and $100 million have accounted for between 16 percent and 17 percent of overall DoD contract obligations in all but one year since 2000 (15 percent in 2003). And contracts obligating between $100 million and $500 million have fluctuated between 14 percent and 18 percent in every year from 2000–2013.

Under sequester, all categories but the largest contracts took significant cuts

Under sequestration, the data indicate that DoD prioritized preserving their biggest acquisition programs, but also indicate that there was no concerted effort to target either smaller contracts or mid-sized contracts as a source of savings. Between 2012 and 2013, obligations in nearly every size category declined at a rate within a percentage point of the overall rate of decline for DoD contract obligations. The only exception was for contracts obligating greater than $500 million, which declined more slowly (-10 percent), increasing as a share of overall DoD contract obligations from 22 percent to 24 percent.

Final Thoughts on Contract Pricing, Vehicle, and Size

The CSIS analysis in this chapter reflects selective samples of data and assessments. Additional data, including breakdowns for all three contract characteristics by major DoD component and area, are available at http://www.csis.org/NSPIR/DoD. CSIS will update both the data and the analysis in response to new policy developments, or as additional data analysis capabilities become available.

7. Methodology of the Study

The Center for Strategic and International Studies (CSIS) has studied and reported on federal and national security contract spending for a decade. This chapter describes the methodology used in this report.

For the purpose of this study, the U.S. defense industrial base is defined as all vendors and individuals that are awarded contracts by the U.S. Department of Defense (DoD). This includes contracts for products, services, and research and development (R&D), classified with the federal supply classification (FSC) codes (also referred to as product or service codes, PSCs).

Most of the data used for this study were derived from the Federal Procurement Data System–Next Generation (FPDS). FPDS covers all federal contract actions that have been awarded during a particular year, although this study is limited to contracts managed by DoD between fiscal years 2000 and 2013. Notably, this approach excludes some contracts funded by DoD but managed by other agencies, because this report focuses on the DoD's acquisition decisions rather than its budget (with the exception of Figure 1-1 and Figure 1-2).

Inherent Restrictions of FPDS

Since the analysis presented in this report relies almost exclusively on FPDS data, it incurs four notable restrictions.

- First, contracts awarded as part of overseas contingency operations are not separately classified in FPDS. As a result, we do not distinguish between contracts funded by base budgets and those funded by supplemental appropriations.
- Second, FPDS includes only prime contracts, and the separate subcontract database has historically been radically incomplete, accounting for less than half of the expected obligations. Therefore, only prime contract data are included in this report.
- Third, reporting regulations require that only unclassified contracts be included in FPDS. We interpret this to mean that few, if any, classified contracts are in the database. For DoD, this omits a substantial amount of total contract spending, perhaps as much as 10 percent. Such omissions are probably most noticeable in R&D contracts.
- Finally, classifications of contracts differ between FPDS and individual vendors. For example, some contracts that a vendor may consider as services are labeled as products in FPDS and vice versa. This may cause some discrepancies between vendors' reports and those of the federal government.

Constant Dollars and Fiscal Years

All dollar amounts in this report are reported as constant fiscal year 2013 dollars unless specifically noted otherwise. Dollar amounts for all years are deflated by the implicit GDP deflator calculated by the

U.S. Bureau of Economic Analysis, with FY 2013 as the base year, allowing the CSIS team to more accurately compare and analyze changes in spending across time. Similarly, all compound annual growth values and percentage growth comparisons are based on constant dollars and thus adjusted for inflation.

Due to the native format of FPDS and the ease of comparison with government databases, all references to years conform to the federal fiscal year. Fiscal year 2013, the most recent complete year in the database, spans October 1, 2012, to September 30, 2013.

Supply Side Classification

Small, Medium, and Large Vendors

To analyze the breakdown of competitors in the market into small, medium, and large vendors, the CSIS team assigned each vendor in the database to one of these size categories. Any organization designated as small by the FPDS database—according to the criteria established by the federal government—was categorized as such unless the vendor was a known subsidiary of a larger entity. Due to varying standards across sectors, an organization may meet the criteria for being a small business in certain contract actions and not in others. The study team did not override these inconsistent entries when calculating the distribution of value by vendor size.

Vendors with annual revenue of more than $3 billion, including from nonfederal sources, are classified as large. This classification is based on the vendor's most recent revenue figure at time of classification. For vendors that have gone out of business or been acquired, this date may be well before 2013. A joint venture between two or more organizations is treated as a single separate entity, and organizations with a large parent are also defined as large. Due to their system integrator role and consistent market share, the study team placed the six largest defense contractors (Lockheed Martin, Boeing, Raytheon, Northrop Grumman, General Dynamics, and BAE) into a separate category called "Big 6 defense vendors." Any vendor assigned a unique identifier by FPDS but is neither small nor large is classified as "medium."

In order to identify large vendors, the study team investigated any vendor with total obligations of $500 million in a single year or $2 billion over the study period. Determining revenues is the most labor-intensive part of the process and involves the use of vendor websites, news articles, various databases, and public financial documents. When taken together, all of this work explains the increase in the market share of large vendors versus some older editions of this report. While large vendors are, on rare occasions, reassigned into the middle tier, the vast majority of investigations either maintain the status quo or identify small or medium vendors that should be classified as large.

Handling of Subsidiaries and Mergers and Acquisitions

To better analyze the defense industrial base, the study team made significant efforts to consolidate data related to subsidiaries and newly acquired vendors with their parent vendors. This results in, among other things, a parent vendor appearing once on CSIS's top 20 lists rather than being divided between multiple entries. The assignment of subsidiaries and mergers to parent vendor is done on an annual basis, and a merger must be completed by the end of March in order to be consolidated for the fiscal year in question. This enabled the study team to more accurately analyze the Defense industrial base, the number of players in it, and the players' level of activity.

Over the past six years, the study team has applied a systematic approach to vendor roll-ups. FPDS uses hundreds of thousands of nine-digit DUNS (Data Universal Numbering System) codes from Dun and Bradstreet to identify service providers. A salutary benefit of this standardization is that FPDS now provides parent vendor codes. These parent codes track the current ownership of vendors but are not backward looking. Thus, a merger that happened in 2010 would not affect parent assignments in 2000. This prevents the study team from adopting these assignments in their entirety. Building off of the work of our departmental reports, we have now expanded *and lowered* that criterion to $250 million of total contract revenue. We have also added an alternative threshold and investigate every DUNS number with more than $1 billion in obligations between 2000 and 2013, no matter how much they receive in any individual year.

We have reinforced these manual DUNS number assignments with automated assignments based on vendor names. Qualifying for an automated assignment by name requires three criteria: 1) a standardized vendor name that matches with the name of a parent vendor, 2) that name has been matched to the parent vendor by the CSIS or the Parent DUNS number field, and 3) there are no alternative CSIS assignments with that vendor name. This process is not immune to error, but it reduces the risk that a DUNS code is considered large in one year but overlooked in another. As an error-checking mechanism, the study team investigated contradictions by comparing our assignments to those made by Parent DUNS numbers for every DUNS number with $500 million in annual obligations or $2 billion in total obligations.

Demand Side Classification: Contract Characteristics

This study considers a variety of contract characteristics: the contracting component, the type of product or service being procured, the funding mechanism, the contract vehicle, the contract size, and the extent of competition. In several cases, this classification can be derived from a single field of the database, using groupings established by the study team. Characteristics that require multiple fields or introduce other complications are listed below.

Competition

The study team followed the DoD methodology and calculated competition by using two fields: extent of competition (which is preferred for awards) and fair opportunity (which is preferred for most IDVs). Additionally, to better evaluate the rate of "effective competition," the study team categorizes competitively awarded contracts by the number of offers received.

Contract Vehicle

Determining the contract vehicle required classifying both awards and indefinite delivery vehicles (IDVs). While classifying awards is straightforward, classifying IDVs requires the referenced IDV contract type field, which is only available via the FPDS web tool. The study team recreates this field by automatically looking up the referenced parent IDV for each delivery order. When this lookup is unsuccessful, typically because the IDV originated before the study period, the study team relies on tables downloaded from the FPDS web tool. This approach may not exactly match the FPDS web tool results, but it allows for cross-tabulation, enables emulation of the DoD method for calculating competition as discussed below, and removes the discrepancies that result from the use of multiple sources.

Contract Size

For the purposes of this report, a contract refers to either an award with a unique procurement identifier or an IDV with a unique pairing of a delivery order procurement identifier and a referenced IDV procurement identifier. Contracts were classified on the basis of total expenditures for the fiscal year in question. Groupings are in nominal dollars because many regulatory thresholds are not adjusted for inflation; as a result, smaller contracts will be slightly overrepresented in recent years. Unlike some prior reports, de-obligations are excluded rather than being grouped with contracts under $250,000.

Data Reliability Notes and Download Dates

Any analysis based on FPDS information is naturally limited by the quality of the underlying data. Several Government Accountability Office (GAO) studies have highlighted the problems of FPDS (for example, the December 30, 2003, report "Reliability of Federal Procurement Data," and the September 27, 2005, report "Improvements Needed for the Federal Procurement Data System–Next Generation").

In addition, FPDS data from past years are continuously updated over time. While fiscal year 2007 was long closed, over $100 billion worth of entries for that year were modified in 2010. This explains any discrepancies between the data presented in this report and those in previous editions. The study team changes over prior year data when a significant change in topline spending is observed in the updates. Tracking these changes does reduce ease of comparison to past years, but the revisions also enable the report to use the best available data and monitor for abuse of updates.

Despite its flaws, the FPDS is the only comprehensive data source of government contracting activity, and it is more than adequate for any analysis focused on trends and order-of-magnitude comparisons. In order to be transparent about weaknesses in the data, this report consistently describes data that could not be classified due to missing entries or contradictory information as "unlabeled" rather than including it in an "other" category.

The 2013 data used in this report were downloaded in February 2014.

About the Project Director and Authors

David J. Berteau is senior vice president and director of the CSIS National Security Program on Industry and Resources, covering national security plans, policies, programs, budgets, and resources; defense management, contracting, logistics, and acquisition; and national security economics and industrial base issues. Mr. Berteau is also an adjunct professor at Georgetown University and at the Lyndon B. Johnson School of Public Affairs, a director of the Procurement Round Table, and a fellow of both the National Academy of Public Administration and the Robert S. Strauss Center at the University of Texas. Before he joined CSIS full time in 2008, he was director of national defense and homeland security for Clark & Weinstock, director of Syracuse University's National Security Studies Program, and senior vice president at Science Applications International Corporation (SAIC), as well as a nonresident senior associate at CSIS. He served at senior levels in the U.S. Defense Department under four defense secretaries, including four years as principal deputy assistant secretary of defense for production and logistics. Mr. Berteau graduated with a B.A. from Tulane University in 1971 and received his master's degree in 1981 from the LBJ School of Public Affairs at the University of Texas.

Gregory Sanders is a fellow with the National Security Program on Industry and Resources at CSIS, where he manages a research team that analyzes data on U.S. government contract spending and other budget and acquisition issues. He employs data visualization and other ways to use complex data collections to create succinct and innovative tables, charts, and maps. His recent research focuses on contract spending by major government departments, contingency contracting in Iraq and Afghanistan, and European and Asian defense budgets. This work requires management of data from a variety of databases, most notably the Federal Procurement Database System, and extensive cross-referencing of multiple budget data sources. In support of these goals, he employs SQL Server, as well as the statistical programming language R. Sanders holds an M.A. in international studies from the University of Denver and a B.A. in government and politics, as well as a B.S. in computer science, from the University of Maryland.

Jesse Ellman is a research associate with the National Security Program on Industry and Resources at CSIS. He specializes in U.S. defense acquisition policy, with a particular focus on recent U.S. Army modernization efforts, federal government cost-estimation methodologies, contract audit/management issues, and federal government contracting policy. Mr. Ellman holds a B.A. in political science from Stony Brook University, and a M.A. with honors in security studies, with a concentration in military operations, from Georgetown University.

Rhys McCormick is a research assistant with the National Security Program on Industry and Resources (NSPIR) at CSIS. His work focuses on unmanned systems, global defense industrial base issues, and U.S. federal and defense contracting trends. Prior to working at NSPIR, he interned at the Abshire-Inamori Leadership Academy at CSIS and the Peacekeeping and Stability Operations Institute at the U.S. Army War College. He holds a B.S. in security and risk analysis from the Pennsylvania State University with concentrations in both intelligence analysis and modeling and information and cyber security.

www.ingramcontent.com/pod-product-compliance
Ingram Content Group UK Ltd.
Pitfield, Milton Keynes, MK11 3LW, UK
UKHW051418150926
475561UK00005B/232